WORLD BOOK

Asteroids, Comets, AND Meteors

Contents

Asteroids

Mete

Comets

If a word is printed in **bold letters that look like this** the first time it appears on any page, you will find the word's meaning in the glossary beginning on page 60.

Astronomers use different kinds of photos to learn about such objects in space as planets. Many photos show an object's natural color. Other photos add false colors or show types of light that the human eye cannot normally see. When appropriate, the captions in this book will state whether a photo uses false color. Other photos and illustrations use color to highlight certain features of interest.

ors

The Smaller Residents
of the Solar System

Most everyone is familiar with the **planets** of our **solar system.** Many can be seen in the night sky, even without using a **telescope.** But the solar system also has many smaller residents. These aren't planets, but they are no less impressive. They include **asteroids, comets,** and **meteors.**

Asteroids and comets—and the meteors that come from them—are objects left over from the formation of our solar system around 4.6 billion years ago.

While the planets have changed over millions of years **orbiting** the sun, many of these small chunks of ice, rock, and metal have not changed much. They are like a fossil record of the solar system!

Our rocky neighbors in the solar system include asteroids similar to these depicted in this artist's drawing.

FUN FACT

In 1971, an astronomer **named an asteroid after his cat,** Mr. Spock (who was named after the fictional character from the popular television show "Star Trek"). The name has stuck, but the International Astronomical Union, which oversees the naming of heavenly bodies, now considers pet names to be unsuitable for asteroids. Naming asteroids for fictional characters is also discouraged.

Many asteroids are named for figures in the myths of ancient Greece and Rome. However, people who discover new asteroids can name the objects after most anyone— such as famous scientists, actors, authors, astronauts, musicians, or teachers.

Asteroids
Not Quite Planets

Most asteroids orbit the sun far from Earth. They shine with reflected sunlight, so they may resemble stars in the night sky. In fact, the term *asteroid* means *starlike.* In reality, asteroids are nothing like stars. They are irregularly shaped rocky bodies.

Astronomers estimate that there are millions of asteroids in our solar system. Most are quite small. Asteroids can range in size from about 600 miles (965 kilometers) in **diameter** to less than 20 feet (6 meters) across. Like planets, asteroids rotate as they orbit the sun.

Asteroids are generally too small to be considered planets. A planet is large enough that the pull of its own **gravity** shapes it into a **sphere.**

Some of the largest asteroids, such as Vesta and Pallas, are almost like planets. They are shaped much like a sphere. Vesta even has a layered internal structure similar to that of Earth and the **moon.** Like Earth and the moon, Vesta has a **crust,** a **mantle,** and an iron **core.** Smaller asteroids usually have a simple structure of rocky or metallic material throughout.

Too small to be a planet, a lone asteroid drifts through the solar system in this artist's illustration.

FUN FACT

Like a planet,

an asteroid can
have a moon!

The asteroid Ida is large enough
that it holds a small moon in
orbit with its gravitational
pull. Ida's tiny moon is named
Dactyl. This tiny moon may
have been a piece of Ida that
broke away in a collision
with another asteroid.

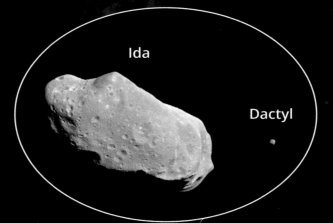

Ida

Dactyl

What Asteroids
Look Like

Some asteroids are dark in color. Others
are bright because they reflect much of

In 1996, the Hubble Space Telescope observed a large impact **crater** on the asteroid Vesta. The crater probably resulted from a collision with another asteroid. Occasionally, a large asteroid will break apart into many smaller ones, usually as the result of a collision. Smaller asteroids are far more common than larger ones.

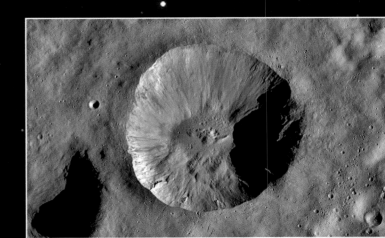

Asteroid
Shapes

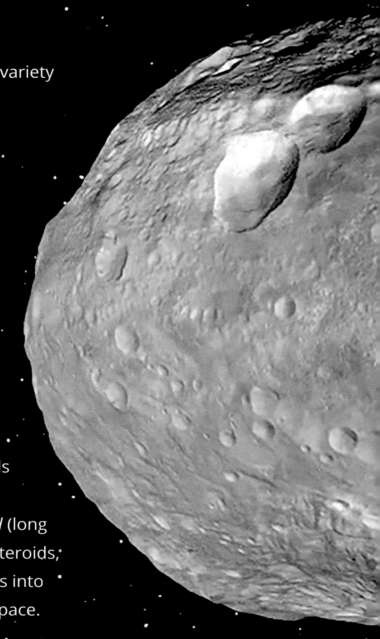

steroids have different surfaces and a variety
f shapes. The largest asteroids appear
oughly spherical because their own
ravitation pulls them into a ball shape.

esta is one such large asteroid. It
measures an average of about 330
miles (530 kilometers) in diameter.
is the only asteroid that can be
een without a telescope—but only
when Vesta is in a favorable position
dark skies, and with knowledge of
where to look. Vesta has the greatest
mass of any known asteroid.

maller asteroids have gravitational pulls
that are too weak to make them round.
hey tend to take on irregular, *elongated* (long
nd thin) forms. Collisions with other asteroids,
rge and small, also shape these objects into
regular chunks of material drifting in space.

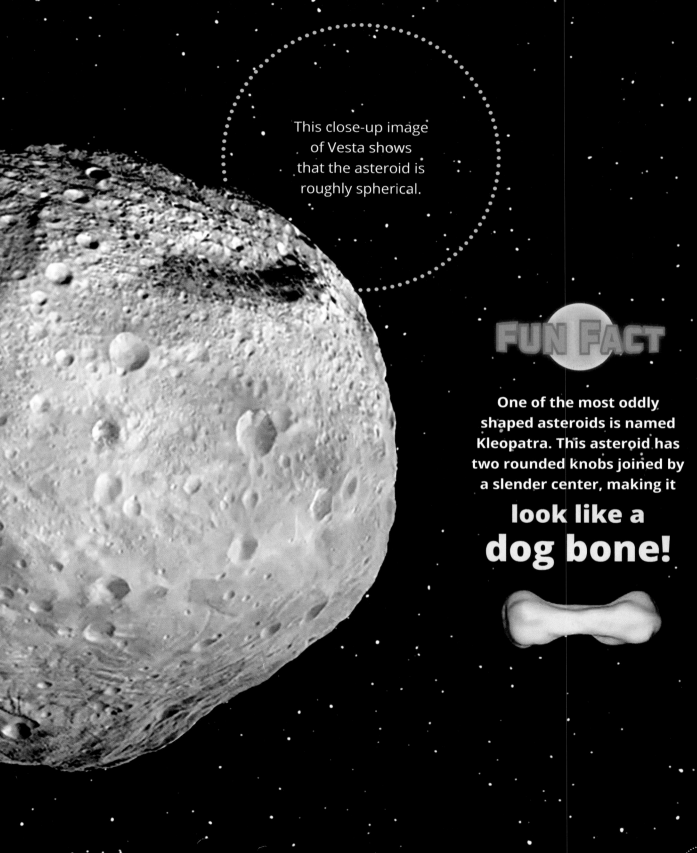

This close-up image of Vesta shows that the asteroid is roughly spherical.

FUN FACT

One of the most oddly shaped asteroids is named Kleopatra. This asteroid has two rounded knobs joined by a slender center, making it **look like a dog bone!**

The Main
Asteroid Belt

The vast majority of asteroids in our solar system are found in a group that astronomers call the **main asteroid belt.** Vesta and Pallas are both in the main asteroid belt.

Main belt asteroids are located in space between the orbits of the planets Mars and Jupiter. The back-and-forth gravitational pull of Jupiter and Mars probably kept the many asteroid pieces from coming together to form a full-sized planet.

The outer part of the main asteroid belt contains many asteroids that are rich in carbon. These asteroids are very old and do not appear to have changed much since the solar system formed about 4.6 billion years ago. The asteroids in the inner part of the belt, which is closer to Earth, contain many metal-rich minerals.

Other groups of asteroids lie farther away from the sun and Earth in the outer regions of our solar system. Two groups known as the *Trojan* asteroids are found in roughly the same orbit as the planet Jupiter. Another group of asteroids, called the *Centaurs,* have orbits that lie between those of Jupiter and Neptune.

Trojan asteroids

More than one million asteroids in the main asteroid belt orbit the sun in the space between Mars and Jupiter.

Asteroid
Composition

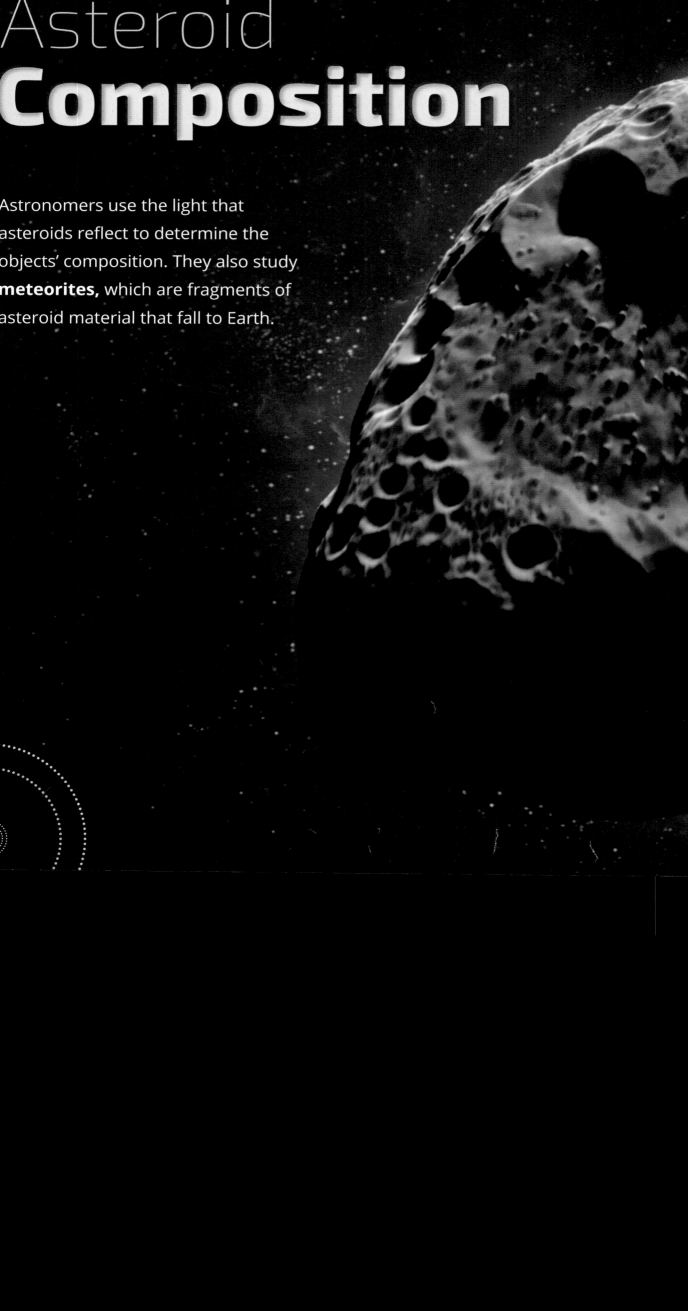

Astronomers use the light that asteroids reflect to determine the objects' composition. They also study **meteorites,** which are fragments of asteroid material that fall to Earth.

Astronomers classify asteroids into one of three major types—*C-type, S-type,* and *M-type.*

C-type asteroids are the most common. They make up more than 70 percent of all asteroids in our solar system. The *C* stands for *chondrite,* a rocky material that is dark in appearance. These are among the most ancient objects in the solar system. They are leftover remnants from the formation of the solar system.

S-type asteroids are stony. They are made up mostly of rocky material with some metals, especially nickel and iron.

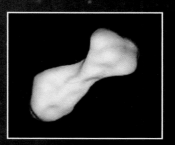

M-type asteroids consist of nearly pure nickel and iron.

Near-Earth
Asteroids

Although most asteroids are found in the main asteroid belt between the orbits of Mars and Jupiter, some are found much closer to Earth. These are known as *near-Earth asteroids.* These objects have orbits that pass close by the orbit of Earth. Some asteroids actually cross Earth's orbital path around the sun. These asteroids could possibly collide with Earth.

Astronomers classify most near-Earth asteroids into one of four groups, called *Apollos, Amors, Atens,* and *Atiras,* based on the size and shape of their orbit compared to the orbit of Earth. The groups are named after a notable asteroid identified in each group.

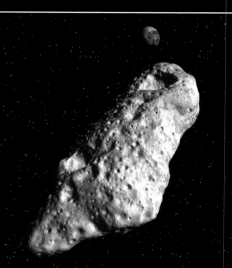

A Close Shave!

In July 2019, an asteroid **larger than a skyscraper** passed by only about 43,000 miles (70,000 kilometers) from Earth. Although there was never any danger of this asteroid striking Earth, astronomers discovered it only a few hours before it zoomed by!

Astronomers have spotted more than 10,000 near-Earth asteroids. Almost 1,000 of these have a diameter of 0.6 miles (1 kilometer) or more. These asteroids could pose a serious threat to Earth. They would cause great destruction if they were to strike our home planet.

Asteroid
Preparedness

Asteroids are not usually a danger to Earth. However, it is possible that a large asteroid could strike Earth. This has happened before. In fact, scientists think that a large asteroid struck Earth about 65 million years ago. The impact caused massive destruction and even caused the **extinction** of the dinosaurs.

When an asteroid strikes the surface of a planet, moon, or even another asteroid, it can produce an *impact crater.* Scientists have found the remains of the impact crater from 65 million years ago in the Gulf of Mexico. Smaller asteroids have struck Earth since the time of the dinosaurs. Some left impact craters that can be seen today.

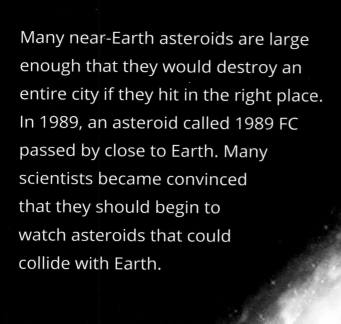

Many near-Earth asteroids are large enough that they would destroy an entire city if they hit in the right place. In 1989, an asteroid called 1989 FC passed by close to Earth. Many scientists became convinced that they should begin to watch asteroids that could collide with Earth.

In 1995, the United States National Aeronautics and Space Administration (NASA) began its Near-Earth Asteroid Tracking (NEAT) program. The scientists in this program use telescopes across the world to look for asteroids that could strike Earth. Scientists are discussing plans of what to do should they discover an asteroid that is on a collision course with Earth.

Studying
Asteroids

In 1991, the space probe

Galileo

took the first close-up pictures of an asteroid called Gaspra. Galileo went on to study the asteroid Ida and discovered its moon, Dactyl.

In 1997, the

Near Earth Asteroid Rendezvous (NEAR)

probe studied the asteroid Mathilde and found many deep impact craters. In February 2000, the NEAR probe made history by going into orbit around the asteroid 433 Eros. That same year, the probe was renamed NEAR-Shoemaker, in honor of the American astronomer Eugene Shoemaker. In 2001, NEAR-Shoemaker made history again when it touched down on the surface of Eros.

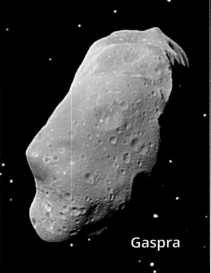

Gaspra

Before 1991, the only way scientists could study asteroids was by using telescopes from Earth. Since then, several space missions have sent **probes** to study asteroids in our solar system.

In 2005, the Japanese spacecraft

Hayabusa

landed on the near-Earth asteroid Itokawa. In 2010, Hayabusa returned to Earth with a small amount of dust taken from this asteroid to be studied by scientists.

NASA's

Dawn

spacecraft, launched in 2007, orbited and photographed Vesta for over a year.

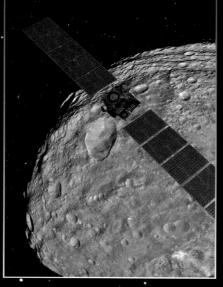

In 2016, NASA launched a space probe called

OSIRIS-Rex

to visit asteroid Bennu. The probe will land on and collect a sample from Bennu and return it to Earth for study in 2023.

Hayabusa2
Shooting an Asteroid

Japan's Hayabusa2 probe was launched in 2014. In 2018, it entered orbit around Ryugu, a near-Earth asteroid that is about 0.6 mile (1 kilometer) in diameter. The spacecraft sent two small **rovers** to land on the surface of Ryugu. The rovers collected information about the asteroid's temperature and composition and sent images to scientists back on Earth.

In early 2019, the orbiting Hayabusa2 probe shot a hard, bullet-like *projectile* into the asteroid. The impact made a small crater on the surface of the asteroid, exposing material underneath. Hayabusa2 then touched down on the surface of Ryugu.

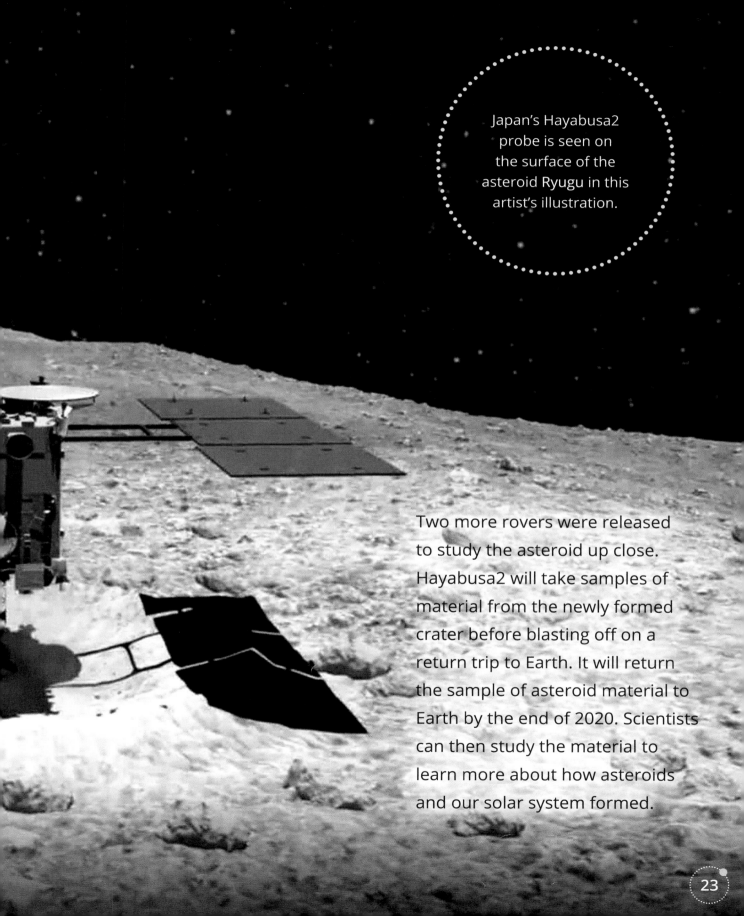

Japan's Hayabusa2 probe is seen on the surface of the asteroid Ryugu in this artist's illustration.

Two more rovers were released to study the asteroid up close. Hayabusa2 will take samples of material from the newly formed crater before blasting off on a return trip to Earth. It will return the sample of asteroid material to Earth by the end of 2020. Scientists can then study the material to learn more about how asteroids and our solar system formed.

Interstellar
Visitor

In 2017, astronomers at the Panoramic Survey Telescope and Rapid Response System (Pan-STARRS) **observatory** in Hawaii spotted an unusual asteroid. Their job is to identify and map the many asteroids that populate our solar system. However, the scientists quickly realized that this object was not a typical asteroid!

The object, officially designated A/2017 UI, passed through our solar system from almost directly above. The object was moving incredibly fast after its closest approach to the sun, about 196,000 miles per hour (355,431 kilometers per hour)—too fast to be captured into orbit around the sun. Astronomers determined that it was highly elongated in shape and rotating rapidly.

Additional data helped confirm that this object formed around a star beyond our own solar system. It was the first such **interstellar** asteroid ever observed. Astronomers at Pan-STARRS named the object 'Oumuamua *(oh oo MOO ah moo ah),* which means *messenger from afar arriving first* in the Hawaiian language.

An artist's
illustration of
'Oumuamua

As 'Oumuamua passed, the sun's gravity caused it to change direction and speed up. The alien visitor then exited the solar system, never to return. The discovery of 'Oumuamua suggests that there are many such interstellar asteroids and they may pass through our solar system frequently. Later observations confirmed

Comets

A comet can be a spectacular sight in the night sky. In ancient times, comets inspired awe and feelings of dread. Ancient people saw these unpredictable visitors as *harbingers* (foretellers) of impending doom and destruction.

In reality, a comet is an icy object that orbits the sun much like planets and asteroids. Astronomers believe that comets are large clumps of ice, rocks, gas, and dust left over from the formation of the outer planets some 4.6 billion years ago. A comet is like a huge, dirty snowball!

The orbit of a comet is usually greatly elongated. As a comet gets closer to the sun, some of the ice starts to melt and boil off, along with particles of dust, sometimes creating a dazzling show in the night sky.

FUN FACT

Rather than bringing doom, some scientists think that comets striking the early Earth brought some of the water and the **organic molecules** that make up the

building blocks of living things!

The Heart of a Comet

A comet is made of a small, solid core. The core is surrounded by a *coma (KOH muh),* a cloud of gas and dust that forms as the sun heats the core. Dust also streams away from the comet in a long tail as it orbits the sun.

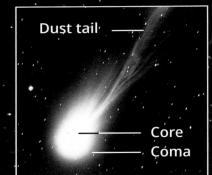

Dust tail

Core

Coma

The core of a comet, also called the **nucleus,** is a ball of ice and rocky dust particles. It also resembles a big, dirty snowball. The ice consists mainly of frozen water. But it may include other frozen chemicals, such as ammonia, carbon dioxide, carbon monoxide, and methane.

The cores of most comets have a diameter of about 10 miles (16 kilometers) or less.

As a comet nears the inner solar system, heat from the sun melts some of the ice on the surface of the core. The core spews gas, water droplets, and dust particles into space. This material forms the comet's coma. **Radiation** from the sun pushes particles away from the coma. These particles form a tail called the *dust tail.*

Most comets are too small to see without a telescope. But some comets become easily visible from Earth when they pass close to the sun. Then, gas and dust in the comet's coma and tail reflect sunlight and become brighter. Some bright comets are even visible in the sky during the daytime!

FUN FACT

The comas of some comets measure nearly 1 million miles (1,6 million kilometers) across. That is 11 times as wide as the diameter of Jupiter, the biggest planet in the solar system.

The longest comet tail recorded

was about 350 million miles (563 million kilometers) long!

Comet McNaught showed off its magnificent tail over Australia in 2007.

Astronomers discovered a second interstellar object zooming past our sun in 2019 when comet 2I/Borisov visited from an unknown distant solar system!

Regular
Visitors

Astronomers classify comets according to how long they take to orbit the sun. *Short-period comets* need less than 200 years to complete one orbit. *Long-period comets* take 200 years or longer to complete one orbit around the sun.

Comets shed gas, ice, and dust each time they return to the inner solar system, leaving behind trails of debris in space. When Earth passes through one of these trails, the particles become meteors that burn up in the **atmosphere.** After coming close to the sun over the course of many orbits, comets eventually lose all their ice. They break up and dissipate into clouds of dust or turn into fragile asteroids.

Long-period comets come from the far reaches of our solar system. Some take a very long time to orbit the sun. For example, Comet Hale-Bopp, which passed by the sun and was visible from Earth in 1997, isn't expected to return until about the year 4380. It had previously passed through the solar system around 2100 B.C.

Comet Hale-Bopp
in the sky over
Croatia in 1997

The Kuiper Belt

Scientists think that short-period comets come from an area at the edge of the solar system called the **Kuiper belt.** The area is named after the Dutch-born American astronomer Gerard P. Kuiper, who described it in 1951. The Kuiper belt begins just beyond the orbit of Neptune, our most distant planet. Thousands of icy clumps drift in this region of the outer solar system. Most are dark and cannot be seen even with powerful telescopes.

FUN FACT

Most comets orbit a safe distance from the sun—Halley's Comet never gets closer than about 55 million miles (89 million kilometers) to the sun. But some comets, called *sungrazers*,

crash straight into the sun

or get so close that they break up, melt, and evaporate.

The comets occasionally are nudged by the gravitational pull from one of the giant planets of the outer solar system—Jupiter, Saturn, Uranus, or Neptune—altering their orbital path. The altered orbit sends them hurtling toward the inner solar system and sun.

Millions of chunks of ice and rock drift in the Kuiper belt.

The Oort Cloud

Long-period comets come from the distant edge of the solar system called the *Oort cloud.* This region of space was named for the Dutch astronomer Jan H. Oort, who first proposed the idea in 1950.

The Oort cloud is like an enormous shell made up of drifting comets surrounding our solar system. It probably begins about 500 billion miles (800 billion kilometers) from the sun. It may extend up to 18 trillion miles (30 trillion kilometers) into space. Scientists think there may be as many as 1 trillion comets in the Oort cloud. Even with so many comets, this region is so vast that most of the icy bodies are separated by thousands of miles or kilometers of empty space.

The gravitational pull of a passing star occasionally disturbs the orbit of comets in the Oort cloud. This may send one into a new orbit that takes it toward the inner solar system and sun as a long-period comet.

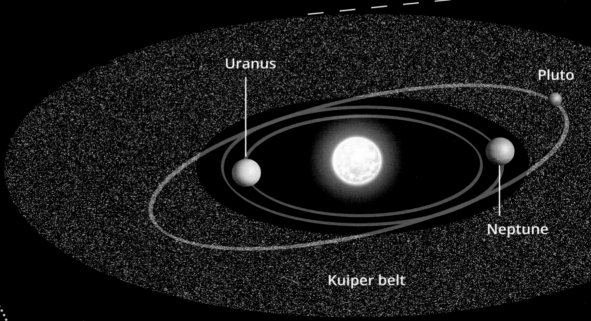

Uranus

Pluto

Neptune

Kuiper belt

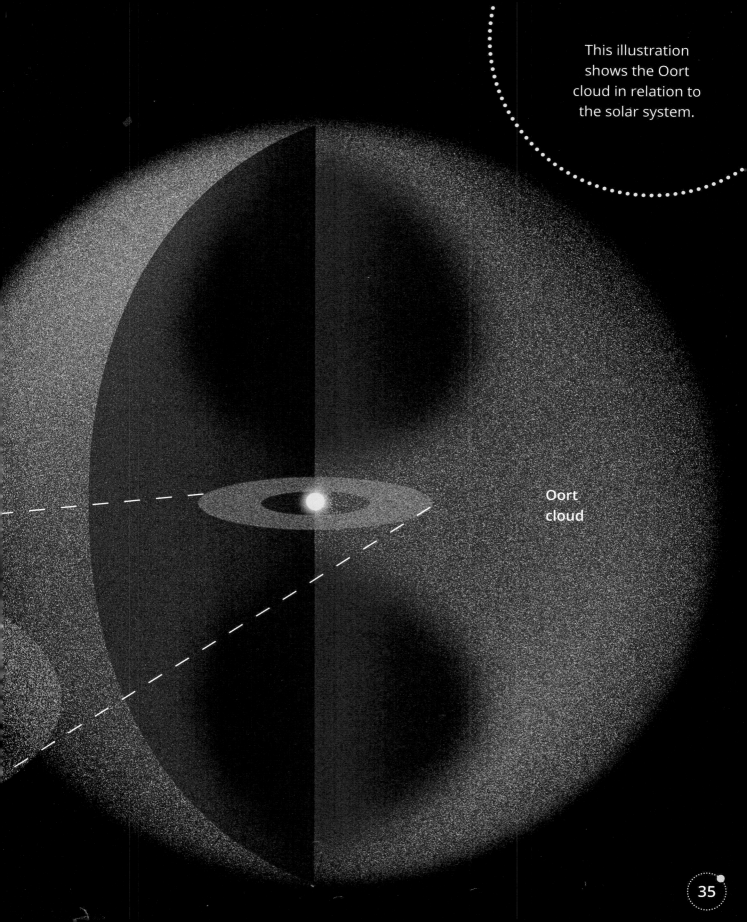

This illustration shows the Oort cloud in relation to the solar system.

Oort cloud

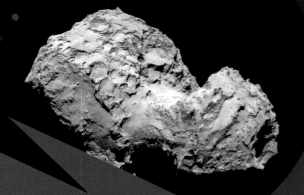

Famous **Comets**

"In the year 1456 ... a Comet was seen passing Retrograde between the Earth and the sun... Hence I dare venture to foretell, that it will return again in the year 1758."

— Edmond Halley

Comet 67P/Churyumov-Gerasimenko

First seen:
1969
Period of orbit (years):
6.45

Comet Hyakutake

First seen:
1996
Period of orbit (years):
63,400

Halley's Comet

First seen:
About 240 B.C.
Period of orbit (years):
76

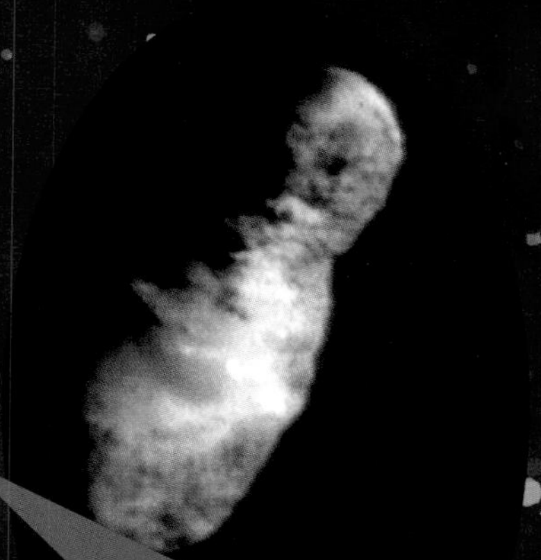

Comet Swift-Tuttle

First seen:
69 B.C.
Period of orbit (years):
130

Comet Borrelly

First seen:
1904
Period of orbit (years):
6.91

Encke's Comet

First seen:
1786
Period of orbit (years):
3.3

Comet Hale-Bopp

First seen:
1995
Period of orbit (years):
2,380

Comet West

First seen:
1975
Period of orbit (years):
558,300

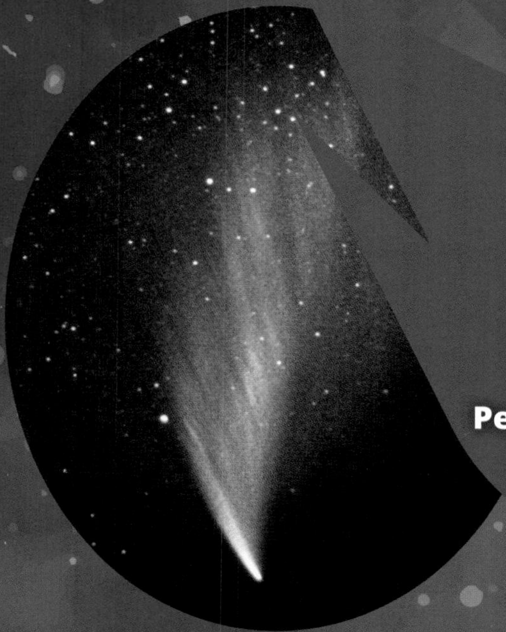

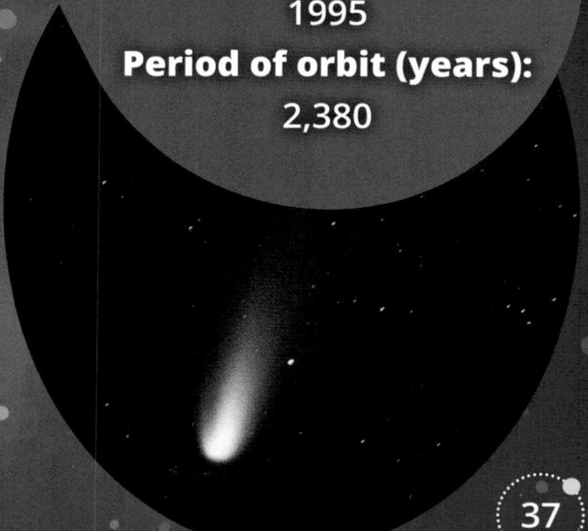

Exploring
Comets

There are many beautiful images taken from Earth of comets streaking through the sky. But astronomers have always wanted to take a closer look at a comet. The best way to do that is with a space probe. In the past few decades, several space probes launched from Earth have approached and photographed comets close up.

Deep Space 1

In 2001, NASA's Deep Space 1 spacecraft flew by comet Borrelly and took detailed photographs of the comet core.

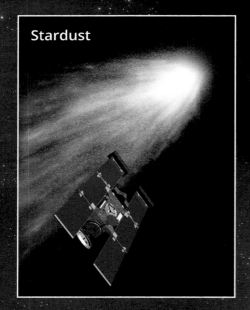

Stardust

NASA's next space probe, called Stardust, flew by a comet named Wild 2 in 2004. Stardust flew within the comet's tail of dust, ice, and debris and returned samples of this material to Earth for scientists to study. Close-up photographs taken by Stardust show the surface of the comet core is rugged, with ice and dust spewing from different spots as the comet approached the sun.

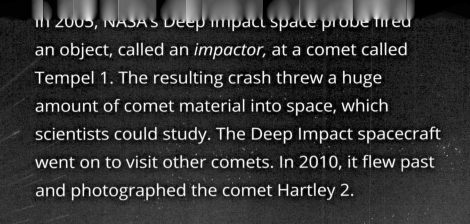

In 2005, NASA's Deep Impact space probe fired an object, called an *impactor*, at a comet called Tempel 1. The resulting crash threw a huge amount of comet material into space, which scientists could study. The Deep Impact spacecraft went on to visit other comets. In 2010, it flew past and photographed the comet Hartley 2.

NASA's Deep Impact probe flies over comet Tempel 1 in this illustration showing the moments just after the probe fired an impactor at the comet.

Landing
on a Comet

How cool would it be to land on a comet? In 2004, the European Space Agency (ESA) launched the Rosetta spacecraft to do just that!

Rosetta began orbiting the comet named 67P/Churyumov-Gerasimenko in 2014. It took 10 years of travel and maneuvering in space to bring Rosetta into orbit around the small comet.

Rosetta carried a small landing craft, called Philae *(FY lee)*, that it launched toward the comet. Philae touched down on the comet surface in November 2014.

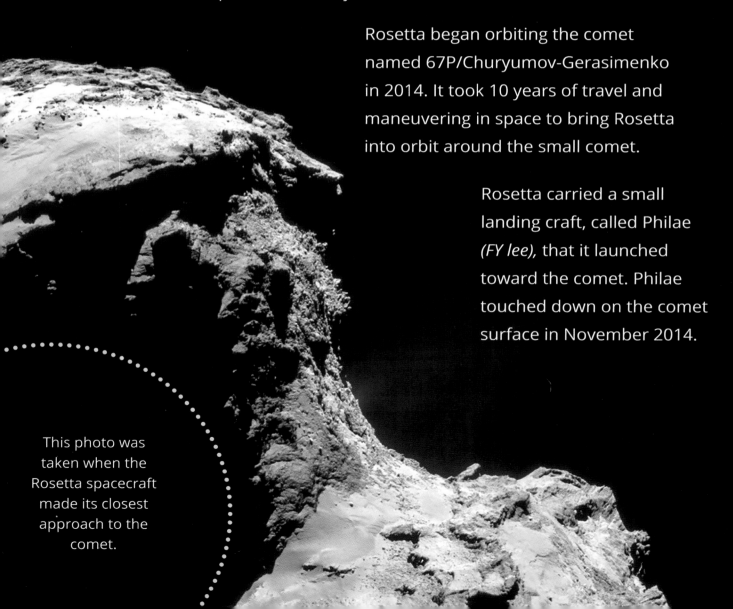

This photo was taken when the Rosetta spacecraft made its closest approach to the comet.

The Rosetta craft recorded Philae's journey (left) to the surface of comet 67P/Churyumov-Gerasimenko.

The lander had a couple of unexpected bounces on the surface before settling slightly off target in a shaded area on the comet core. The lander sent several close-up photographs of the comet surface. These showed a bleak landscape of rubble and ice.

The Philae la
captured this
of a cliff on
surface of co
67P/Churyur
Gerasimen
Astronomers r
it Perihelion

In the shade, without sunlight reaching its solar panels, Philae could not recharge its batteries. The lander could only take photographs and record information for about 2 ½ days before the batteries were drained, ending the mission.

In September 2016, as the comet headed back toward the outer solar system, ESA scientists guided Rosetta to crash into it, ending its mission. The probe continued to take photographs and gather data until just seconds before the collision.

This image, made from a series of photos, shows the fragments of comet Shoemaker-Levy 9 speeding toward a collision with Jupiter.

Comet Impact!

In 1993, the American astronomers Carolyn and Eugene Shoemaker and the Canadian-born astronomer David H. Levy discovered a new comet. They quickly calculated that the orbit of this comet would have it pass very near the planet Jupiter.

But the astronomers soon realized that the comet, named Shoemaker-Levy 9, had been captured by Jupiter's powerful gravitational pull! The comet was going to crash into the giant planet in a huge cosmic collision!

As Shoemaker-Levy 9 neared Jupiter, the planet's gravitational pull broke the icy body into more than 20 fragments. Astronomers from around the world watched as most of the fragments smashed into Jupiter over a period of several days in July 1994. It was the first time astronomers had ever witnessed the collision of a comet and a planet!

Although all of the comet fragments collided on the side of Jupiter facing away from Earth, each impact caused a huge explosion in the upper atmosphere of Jupiter. The scars from each impact remained visible from Earth for more than a month.

NASA's Galileo space probe captured these images of comet fragments smashing into Jupiter.

Comets in
History and Folklore

A comet in the night sky is an awe-inspiring sight. But, throughout history, people have viewed comets with fear.

Unlike the **constellations** and planets, which move across the night sky in an orderly parade, comets are unpredictable visitors. Many people in ancient times took the sudden appearance of a dazzling comet in the sky as a bad omen or warning from the gods.

The arrival of a comet was almost always associated with bad news—a natural disaster, the death of a king, crop failure,

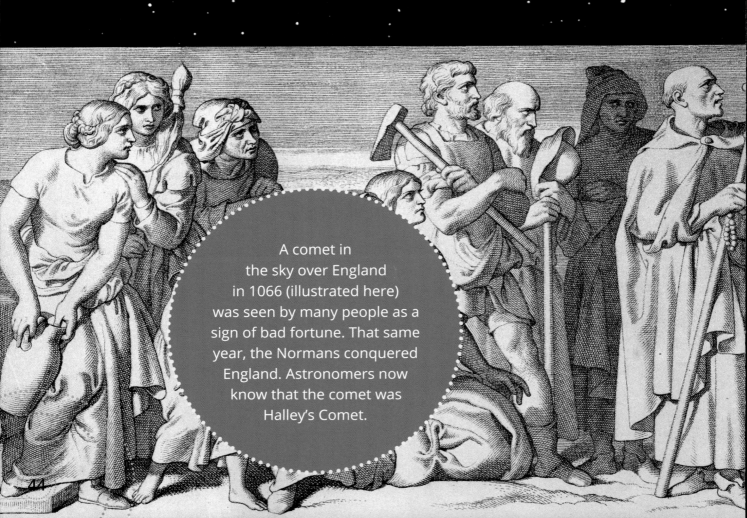

A comet in the sky over England in 1066 (illustrated here) was seen by many people as a sign of bad fortune. That same year, the Normans conquered England. Astronomers now know that the comet was Halley's Comet.

or an outbreak of disease. Legend has it that in ancient Rome, astronomers observed a brilliant comet that arrived in the sky during funeral ceremonies following the death of the Roman emperor Julius Caesar. They took this as proof that Caesar was *divine* (godlike). According to another legend, Inca astronomers in South America sighted a comet not long before the Spanish conqueror Francisco Pizarro arrived to cause the downfall of their civilization.

Astronomers in ancient China paid great attention to the arrival of comets in the night sky. For centuries, they kept careful records on the arrival of comets, their position in the sky, the appearance and direction of travel, and when they disappeared from view. The detailed records of comets by Chinese astronomers helped modern astronomers understand the nature of comets.

Meteors
Visitors from Space

According to popular folklore, if you see a shooting star, you should make a wish! People often call meteors *shooting stars* or *falling stars* because they look like stars falling from the sky.

A meteor appears streaking through the sky when a piece of matter enters Earth's atmosphere from space at high speed. Such a piece of matter is called a **meteoroid.** This term only applies when the object is in space.

Meteors are always
colliding with Earth's
atmosphere. Astronomers
estimate that about 48
tons (44,000 kilograms) of
meteorite material
fall on Earth
every day!

As a meteoroid collides with the air in Earth's atmosphere,
it is heated so that it glows—a meteor. Zooming through
the atmosphere at high speed, a meteor may leave a
shining trail of hot gases. Most meteoroids that cause
visible meteors are smaller than a pebble. Most
meteors break apart and disappear from view
within seconds as they streak through Earth's
atmosphere.

If a meteor remains intact and falls to Earth,
the surviving piece is called a **meteorite.**

Types
of Meteorites

Astronomers classify meteorites into groups based on their composition. There are

three
basic kinds

of meteorites: (1) *stony*, (2) *stony-iron*, and (3) *iron*.

Stony meteorites consist of minerals rich in silicon and oxygen. One group of stony meteorites, called *chondrites (KON dryts)*, are made of the same material from which

the planets of the solar system formed.

Another group of stony meteorites, called *achondrites (ay KON dryts)*, were

once part of
an asteroid

large enough to have melted and separated into an iron-rich core and a stony crust. Achondrite meteorites come from the outer crust of these bodies.

Iron meteorites are made mostly of iron and often some nickel. They originate **from the core of larger asteroids.**

Stony-iron meteorites are **made of stone and metal,** usually iron and nickel. They originate from the part of an asteroid between the crust and the core.

Most meteorites originate from asteroids.

Pieces of an asteroid often break free in collisions with other asteroids. Asteroids are made up of stronger material than that of comets. Such material is less likely to break up as it falls through the atmosphere to Earth.

Finding
Meteorites

It can be hard to identify a meteorite. They can be difficult to distinguish from Earth rocks by appearance alone. But, if you find an odd-looking rock, check it out! It may be a meteorite!

Nearly all meteors break apart in Earth's atmosphere. Traveling at thousands of miles or kilometers per hour, the rocky material disintegrates into a bright flare. Only a small number of meteors survive this fiery journey to reach the ground. These meteorites usually range between the size of a pebble and a large apple. However, some very large meteorites have also struck Earth and been found.

In 2008, astronomers spotted a near-Earth asteroid about 13 feet (4 meters) in diameter. They determined that it would collide with Earth's atmosphere as a meteor and calculated where it would hit. As they predicted, the meteor exploded in the sky above Sudan in Africa. Later, they collected more than

600 meteorites,

weighing a total of about 24 pounds (11 kilograms), that landed in the Sudan desert after the explosion.

A dark meteorite is easily seen where it landed on a bright ice field in Antarctica.

Meteorite hunters have a much easier time searching in barren deserts and other open areas. In deserts, dark meteorites can easily be seen when they fall upon light-colored sand. Many meteorites are found in the snow-covered plains of Antarctica.

From the Moon and Mars!

Most meteorites are remnants of asteroids—they are original materials left over from the formation of the solar system billions of years ago. By studying meteorites, scientists can learn about the conditions and processes that formed Earth and other planets.

But scientists have also found a small number of meteorites that originated on the moon. A few other unusual meteorites have even come to Earth from Mars!

Scientists think that an asteroid occasionally crashes into the moon or Mars. If the rock is big enough, the resulting explosion can throw chunks of moon or Martian rock into space. Eventually some of these pieces may drift toward Earth, where they fall through the sky as meteors and land as meteorites.

Scientists can tell that a particular meteorite came from the moon because the chemical composition of the meteorite matches that of moon rocks returned to Earth by astronauts. Scientists can identify a meteorite from Mars from the chemicals present in tiny pockets of gas preserved in the meteorite. The composition of the meteorite gases matches that of gases found in the atmosphere of Mars as measured by orbiting spacecraft and probes.

Scientists have discovered that this meteorite came to Earth from Mars!

1 CM

An artist's illustration of the Tunguska meteor in the sky over Siberia

The Tunguska Event

Occasionally, Earth gets hit by a very large meteor. On June 30, 1908, one such meteor entered Earth's atmosphere and exploded over Tunguska *(toon GOO skuh)*, an isolated region of Russian Siberia. Only a few people witnessed the meteor streaking across the sky because it occurred in a region with little population. The meteor broke apart in a huge explosion about 3 to 6 miles (5 to 10 kilometers) above the ground. The blast flattened trees over the forested region like an atomic bomb!

The Tunguska event was the largest meteor impact on Earth in recorded history. Nobody was harmed in this devastating event, because few people live in this remote region. Perhaps because of the isolation of the area there was relatively little investigation of the blast immediately after it occurred.

In the 1920's, the Russian scientist Leonid Kulik led a scientific party to the region to determine what happened. Widespread damage from the blast was still evident after many years. Kulik concluded that the destruction was likely caused by a meteor exploding high above. The scientists determined that the meteor was destroyed in the blast, since they were unable to find any surviving meteorites on the ground.

Meteor Showers

You can usually see meteors in the sky on most any clear night—especially in dark regions far from city lights. However, at certain times of the year, the rate at which meteors appear in the sky increases dramatically. These are called *meteor showers*.

Meteor showers occur when Earth passes through a stream of meteoroids in space. The meteoroids are often dust and other particles that were thrown off and left to drift in space by a passing comet. Since Earth orbits the sun, it passes through this debris around the same time each year. Seen from the ground, all the meteors in a shower appear to come from the same direction in the sky.

Struck by a Meteorite!

Even though thousands of meteors rain down on Earth, the danger to people is incredibly small. In 1954, a grapefruit-size meteorite smashed through the roof of a house in Sylacauga, Alabama. The space rock crashed through the ceiling, bounced off a radio, and hit a woman as she lay in bed. The woman suffered only a bruise. This remains the only known instance in history of a person being struck by a meteorite!

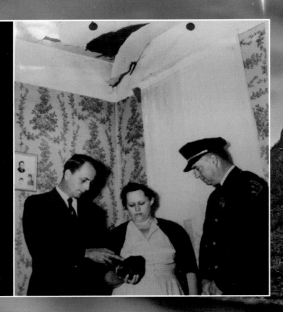

The meteors in a shower appear to come from the same direction in the sky.

One of the most well-known annual meteor showers is called the *Perseid (PUR see ihd)* meteor shower. It occurs in the middle of August. The meteors appear to come from a region of the sky near the constellation Perseus. At its peak, 50 to 100 meteors may be seen streaking through the sky every hour. Every Perseid meteor is a tiny piece of a comet named Swift-Tuttle, which orbits the sun every 135 years.

Over the span of a meteor shower, the rate at which meteors appear increases, peaks, and then decreases as Earth moves in and then out of a stream. If the rate of meteors appearing is very high—hundreds per hour, it is called a *meteor storm.*

Russia Fireball

On Feb. 15, 2013, a large meteor burned brightly through the sky, creating a *fireball,* above the town of Chelyabinsk *(chel YAH binsk),* in southeastern Russia.

Scientists think the chunk of rock that caused this meteor was about 65 feet (20 meters) across and weighed more than 11,000 tons (11 million kilograms)! They calculate that the space rock was traveling at about 12 miles (19 kilometers) per second when it first struck Earth's atmosphere.

The meteor exploded about 15 miles (25 kilometers) above the ground. Nobody was killed in the blast.

However, the blast shattered windows over a large area and damaged many buildings. About 1,500 people were injured, mostly by flying glass from shattered windows.

Several meteorites landed on the ground in this event. One large meteorite punched a 20-foot (6-meter) hole in the ice covering a small lake southwest of the town. The fragment weighed over 1,300 pounds (600 kilograms).

The Chelyabinsk meteor was the largest seen since the meteor that exploded over the Tunguska region of Siberia in 1908.

Glossary

asteroid A small body made of rocky material or metal that orbits a star.

astronomer A scientist who studies stars, planets, and other objects or forces in space.

atmosphere *(AT muh sfihr)* The mass of gases that surrounds a planet or other body.

carbon A common chemical element that is black in color. Carbon occurs in combination with other elements in all plants and animals.

comet A small body made of dirt and ice that orbits the sun.

constellation A group of stars, usually having a geometric shape within a definite region of the sky. Constellations are often named after mythological figures.

core The center part of the inside of a planet, moon, or star.

crater A bowl-shaped depression on the surface of a moon or planet created by the impact of an object.

crust The solid, outer layer of a planet, moon, or star.

diameter The length of a straight line through the middle of a circle or anything shaped like a ball.

extinction The complete dying out of a species or group of living things.

gravity The force of attraction that acts between all objects because of their mass.

interstellar Situated or taking place between the stars.

Kuiper *(KY pur)* **belt** A ring of icy objects orbiting in the outer solar system beyond Neptune. Scientists believe that many comets are objects from the Kuiper belt.

main asteroid belt The region between Mars and Jupiter where most asteroids exist.

mantle The area of a planet or moon between the crust and the core.

mass The amount of matter that an object has.

meteor A mass of stone or metal that appears as a streak of light in the sky.

meteorite A mass of stone or metal from outer space that has reached the surface of a planet or moon without burning up in that body's atmosphere.

meteoroid A small object, believed to be the remains of a disintegrated comet, which travels through space.

mineral A substance, such as tin, salt, or sulfur, that is formed naturally in rocks.

molecule The smallest particle into which a substance can be divided without chemical change. A molecule of an element consists of one or more atoms that are alike; a molecule of a compound consists of two or more different atoms.

moon A smaller body that orbits a planet or asteroid.

nucleus The central part, or core, of an object. (The plural is *nuclei.*)

observatory A place or building with a telescope or other equipment for observing the stars and other heavenly bodies.

orbit The path that a smaller body takes around a larger body; for instance, the path that a planet takes around the sun.

organic *(awr GAN ihk)* Any chemical compound containing the element carbon.

planet A large, round body in space that orbits a star. A planet must have sufficient gravitational pull to clear other objects from the area of its orbit.

probe An unpiloted device sent to explore space. Most probes send *data* (information) from space back to Earth.

radiation Energy given off in the form of waves or small particles of matter.

rover A robotic device that can move about, explore, and collect information under its own power via remote control.

solar system A group of bodies in space made up of a star and the planets and other objects orbiting around that star.

sphere A round, ball-shaped object.

telescope An instrument for making distant objects appear nearer and larger. Simple telescopes usually consist of an arrangement of lenses, and sometimes mirrors, in one or more tubes.

Index

World Book, Inc.
180 North LaSalle Street
Suite 900
Chicago, Illinois 60601
USA

For information about other "Solar System" titles, as well as other World Book print and digital publications, please go to www.worldbook.com or call 1-800-WORLDBK (967-5325).

For information about sales to schools and libraries, call 1-800-975-3250 (United States) or 1-800-837-5365 (Canada).

Library of Congress Cataloging-in-Publication Data for this volume has been applied for.

Our Solar System
ISBN: 978-0-7166-8058-1 (set, hc.)

Asteroids, Comets, and Meteors
ISBN: 978-0-7166-8067-3 (hc.)

Also available as:
ISBN: 978-0-7166-8077-2 (e-book)

Printed in the United States of America
by CG Book Printers,
North Mankato, Minnesota
1st printing March 2020

Staff

Editorial

Writer
Nicholas Kilzer

Senior Editor
Shawn Brennan

Editors
Will Adams
Mellonee Carrigan

Proofreader
Nathalie Strassheim

Manager, Indexing Services
David Pofelski

Graphics and Design

Senior Visual Communications Designer
Melanie Bender

Media Editor
Rosalia Bledsoe

Manufacturing/Production

Manufacturing Manager
Anne Fritzinger

Production Specialist
Curley Hunter

Acknowledgments

Cover: © Jurik Peter, Shutterstock; © Digital Storm/Shutterstock
1-5 © Shutterstock
6-7 © Mark Garlick, Science Photo Library/Getty Images; NASA/JPL
8-9 © Nostalgia for Infinity/Shutterstock; © Dimpank/Shutterstock; NASA/JPL-Caltech/UCLA/MPS/DLR/IDA
10-11 NASA/JPL-Caltech/UCLA/MPS/DLR/IDA; NASA/JPL
12-13 © Johan Swanepoel, Shutterstock; NASA/JPL-Caltech
14-15 © Digital Storm/Shutterstock; NASA/JPL/JHUAPL; NASA/JPL
16-19 © Shutterstock
20-21 NASA; NASA/JPL/USGS; NASA/JPL-Caltech; German Aerospace Center (licensed under CC BY 3.0)
22-23 JAXA
24-27 © Shutterstock
28-29 © John White Photos/Alamy Images; Lick Observatory
30-31 NASA, ESA, and D. Jewitt (UCLA); Philipp Salzgeber (licensed under CC BY-SA 2.0 AT)
32-33 ESO/M. Kornmesser (licensed under CC BY 2.0)
34-35 © Tim Brown, Science Source
36-37 © Vende Design/Shutterstock; NASA/ESA/Giotto Project; E. Kolmhofer, H. Raab; Johannes-Kepler-Observatory, Linz, Austria (licensed under CC BY-SA 3.0); ESA/Rosetta/MPS for OSIRIS Team; MPS/UPD/LAM/IAA/SSO/INTA/UPM/DASP/IDA; NASA; NASA/JPL; ESO/P. Stättmayer (licensed under CC BY 4.0); NASA/JPL-Caltech/Univ. of Minn.
38-39 NASA/JPL/UMD; NASA/JPL
40-41 ESA/Rosetta/NavCam (CC BY-SA 3.0 IGO); NASA/ESA/Rosetta/OSIRIS/MPS/UPD/LAM/IAA/SSO/INTA/UPM/DASP/IDA; NASA/ESA/Rosetta/Philae/CIVA
42-43 NASA, ESA, H. Weaver and E. Smith STScI and J. Trauger and R. Evans NASA Jet Propulsion Laboratory; NASA/JPL
44-45 © Duncan1890/Getty Images
46-49 © Shutterstock
50-51 NASA/Cindy Evans; NASA/SETI/P. Jenniskens
52-53 © Detlev van Ravenswaay, Science Source; NASA
54-55 © Detlev van Ravenswaay, Science Source; Public Domain
56-57 © Haitong Yu, Getty Images; © The University of Alabama
58-59 M. Ahmetvaleev/NASA/JPL-Caltech